END HUMAN TRAFFICKING

Prayer and Resource Guide

END HUMAN TRAFFICKING

Prayer and Resource Guide

Edited
by
T. A. Riebel, MSW, CCTP

Afterword by Kara Griffin,
Associate lay member, *U.S. Catholic Sisters Against Human Trafficking*

END HUMAN TRAFFICKING Prayer and Resource Guide
First Edition

ISBN 978-0-359-42027-8

Savior of the World, and *Jesus the True Shepherd,* by
Thomas Kelly, 1871-1874. N.Y., [1874] Public Domain

Recreative image of Blessed Mother (1900),
Berg, Charles I., 1856-1926. Public Domain.

Shield in battle [White House] (angel wings), 1880
Johnston, Frances Benjamin, 1864-1952. Public Domain.

Holy Spirit, fire, dove, 1857
J & R Lamb Studios. Public Domain.

Lulu Press
Morrisville, NC, USA
www.lulu.com
1844-212-0689
Purchase this book at discount through the above distributor, or it is
available through retail bookstores.

Printed in the United States of America and worldwide.
U.S. trade-ready print book.

Library of Congress Control Number 2019902065

ISBN 978-0-359-42027-8

Dedication

For the victims of Human Trafficking.

CONTENTS

(Victim-Witness federal protection resources page 57)

Preface

Human trafficking is modern-day slavery that involves the use of force, fraud, or coercion to obtain some type of labor or commercial sex act. (Department of Homeland Security)

There are approximately 20 to 30 million people enslaved in the world today. (CNN Freedom Project)

According to the U.S. State Department: 600,000 to 800,000 people are trafficked across international borders every year.

Who does this happen to?   Men, women, and children; all ages, races, and nations.

Types of Trafficking:

- Forced labor
- Organ harvesting
- Sex trafficking
- Domestic servitude
- Sexual exploitation

Human Trafficking victims are reported to be 55% female, and 45% male. The average age of a teenager in trafficking in the United States is ages 12-14. (hhs.gov)

Between 14,500 to 17,500 people are trafficked into the U.S. each year.

The average cost of a slave is $90. (freetheslaves.net)

Human Trafficking generates approximately $32 billion in human body sales each year. (CNN)

Human Trafficking is the most heinous, grave evil known to the world today.

Introduction

This book, *End Human Trafficking: Prayer and Resource Guide* is a source of prayer and action in the fight to end human trafficking.

National Human Trafficking Awareness Day is dedicated for January 11, and World Day Against Trafficking in Persons (United Nations) is on July 30 each year.

A further day of awareness for this serious cause of human trafficking is on February 8, which is the feast day for Catholic saint, canonized by St. Pope John Paul II in 2000, St. Josephine Bakhita. Among her patronage, is for the concern of Human Trafficking, and February 8 each year is marked as the International Day of Prayer and Awareness for Human Trafficking. Josephine M. Bakhita (1869-1947) was herself sold into slavery and trafficking, and after many years of torture, was freed, and converted to Catholicism. This book thus provides several prayer devotionals, written in the Catholic Christian tradition, as well as faith resources for the cause of human trafficking.

Additionally, are sections in the book devoted to clinical outlines of trafficking, victim symptomology and basic needs. There are explanations on (PTSD) Post-Traumatic Stress Disorder, and Trauma-Informed Care. You will find sections on laws pertaining to the fight against trafficking, professional training for intervention in trafficking, and federal and local resources for the victims of human trafficking. Finally, a Recovery and Reintegration section brings to light important insights and information to know if you are an educator, advocate, volunteer, or service provider to the victims of human trafficking.

On behalf of those victimized, thank you. Thank you for every prayer, thought, outreach, and action to help save these suffering people enslaved as a victim of human trafficking.

Prayers

for

Human Trafficking

How I wish that all of us would hear God's cry:
"Where is your brother?" (Genesis 4:9)
Where is your brother or sister who is enslaved?
Where is the brother and sister who you
are killing each day
in clandestine warehouses,
in rings of prostitution,
in children used for begging,
in exploiting undocumented labour?
Let us not look the other way.

Pope Francis
Envagelii Gaudieum

Litany for the Enslaved

Oh God, our only Master, save your people.
Oh Lord, obliterate this evil.
Oh Spirit, destroy these actions.

For the children taken into slavery, *save them O Lord.*
For the women taken into slavery, *save them O Lord.*
For the men taken into slavery, *save them O Lord.*

For those in forced labor, *save them O Lord.*
For those in domestic servitude, *save them O Lord.*
For those removed from their land, *save them O Lord.*

For those sold for sex, *save them O Lord.*
For those being exploited and abused, *save them O Lord.*
For those being prostituted, *save them O Lord.*

For those killed, *save them O Lord.*
For the organs sold, *save them O Lord.*
For those tortured, *save them O Lord.*

Oh God, our only Master, save your people. *We cry to you.*
Oh Lord, obliterate this evil. *We cry to you.*
Oh Spirit, destroy these actions. *We cry to you.*

We beg You most powerful God, to put an end to human
trafficking now and for all eternity. *Hear us O Lord.*

We plead to You most powerful God, to put an end to human
trafficking now and for all eternity. *Hear us O Lord.*

Oh Lord, destroy this evil. *Hear us Oh Lord.*
Oh, Lord, destroy this evil. *Hear us Oh Lord.*
Oh, Lord, destroy this evil. *Hear us Oh Lord.*

In Jesus name, we command this evil disappear. *Amen.*
In Jesus name, we command this evil disappear. *Amen.*
In Jesus name, we command this evil disappear. *Amen.*

O Spirit, *end this torture.*
O Spirit, *end this torture.*
O Spirit, *end this torture.*

Oh God, our only Master, save your people.
Oh Lord, obliterate this evil.
Oh Spirit, destroy these actions.

For the children taken into slavery, *save them O Lord.*
For the women taken into slavery, *save them O Lord.*
For the men taken into slavery, *save them O Lord.*

For those in forced labor, *save them O Lord.*
For those in domestic servitude, *save them O Lord.*
For those removed from their land, *save them O Lord.*

For those sold for sex, *save them O Lord.*
For those being exploited and abused, *save them O Lord.*
For those being prostituted, *save them O Lord.*

We beg You most powerful God, to put an end to human trafficking now and for all eternity. *Hear us O Lord.*

We plead to You most powerful God, to put an end to human trafficking now and for all eternity. *Hear us O Lord.*

Chaplet to End Trafficking

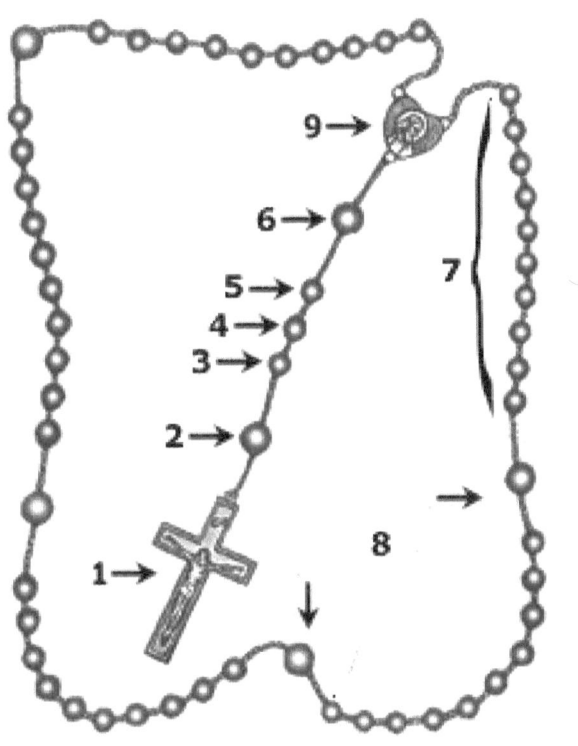

Instructions

1. Sign of the Cross
2. Apostles' Creed
3. *Save them O Lord.*
4. *Save them O Lord.*
5. *Save them O Lord.*
6. Holy Spirit prayer
7. *Jesus, destroy this evil.*
8. Most pure Mother, Divine Sacrifice
9. Most pure Mother, Divine Sacrifice

Apostles' Creed (Catholic Church)

I believe in God,
the Father almighty,
Creator of heaven and earth,
and in Jesus Christ, his only Son, our Lord,
who was conceived by the Holy Spirit,
born of the Virgin Mary,
suffered under Pontius Pilate,
was crucified, died and was buried;
he descended into hell;
on the third day he rose again from the dead;
he ascended into heaven,
and is seated at the right hand of God the Father almighty;
from there he will come to judge the living and the dead.
I believe in the Holy Spirit,
the holy catholic Church,
the communion of saints,
the forgiveness of sins,
the resurrection of the body,
and life everlasting. Amen.

Holy Spirit Prayer (altered/specific to cause)

Come Holy Spirit, fill the hearts of Your faithful,
and enkindle in them the Fire of Your love.
Send forth Your Spirit,
and we shall be created,
and You shall renew the face of the earth.
Holy Spirit, terror of demons, intercede for these victims.

Mother Most Pure, Divine Sacrifice (altered/specific to cause)

Mother most pure,
we offer You the Body and Blood,
Soul and Divinity,
of Your Dearly Beloved Son,
our Lord, Jesus Christ,
this, for an end to human trafficking.

1.   In the name of the Father, and of the Son, and of the Holy Spirit. Amen. **(All)**

1.   Apostles' Creed **(Leader/All)**

3-5. *Save them O Lord.* **(All)**

6.   Come Holy Spirit, fill the hearts of Your faithful, **(Leader)**
     and enkindle in them the fire of Your love.
     send forth Your Spirit,
     and we shall be created,
     and You shall renew the face of the earth.
     Holy Spirit, terror of demons, intercede for these victims.

7.   **[All small beads, 10 each x 5 decades]**

     Jesus, destroy this evil. **(All)**

8.   **[single bead between decades] (Leader)**

     Mother Most Pure, Divine Sacrifice

     Mother most pure,
     we offer You the Body and Blood,
     Soul and Divinity,
     of Your Dearly Beloved Son,
     our Lord, Jesus Christ,
     this, for an end to human trafficking.

9.   **[Closing] (All)**

     Mother Most Pure, Divine Sacrifice

     Mother most pure,
     we offer You the Body and Blood,
     Soul and Divinity,
     of Your Dearly Beloved Son,
     our Lord, Jesus Christ,
     this, for an end to human trafficking.

Plea to Mary, Virgin Mother of God

Most holy and pure Virgin Mother,
there is a grave evil of human trafficking
active all over the world.

Shine your pure light on every evil
of these traffickers,
and let us capture and prosecute these criminals.

Shine your pure light on every evil
of these traffickers,
and let us report every activity that we see around us.

Purify our minds, so that
we may never act or speak in an immoral manner.
Purify our minds, so that
we may never be source of evil.

Most holy and pure Virgin Mother,
pour down Your grace upon these victims,
that they may receive an abundance of help,
and healing from the suffering.

In Jesus' name. Amen

For those involved in prostitution

Dear Lord,

We pray for those involved in the
horrible crimes of prostitution.
These grave immoral actions
are a crime against the body,
and a crime against Your will for our lives.
Please guide those who are working to stop these crimes
in order to save lives and souls.
Put an end to these evils once and for all.
Teach us the proper use of our bodies and sexuality.
Change our minds, change our hearts,
change our attitude.
Correct our thinking that we may know
the difference between right and wrong.
Let us be a representation of Your will for our people,
in all Your goodness and splendor.
We Your people say no to these bad actions.
Dear Mother, purify Your children.
Give these people health of mind and body,
and a change of life.
I offer this intercessory prayer for those
involved in prostitution.

In Jesus' name, Amen.

For those in human trafficking

Dear Lord,

We pray for those involved in the
horrible crimes of human trafficking;
in labor trafficking, sex trafficking, organ trafficking,
and all forms of human trafficking.
These heinous crimes are unthinkable,
and yet they are happening all over our world
to children, teenagers, and adults.
Make me a voice and life of opposition
to this severely grave and deadly activity.
Put an end to these dark evils once and for all.
Trafficking is a grave evil,
and in all opposition to Your plan of love and goodness.
There is nothing in trafficking
that is for Your people or our world.
Please shield and guard these victims,
and send Your army of angels to intervene.
Assist all those who work for this cause,
that they may make every advancement against this evil.
I offer this intercessory prayer for those
involved in human trafficking.

In Jesus' name, Amen.

Stations of the Cross to End Human Trafficking

SAVIOUR OF THE WORLD

Publ. & Print by Th.Kelly 17 Barclay St. N.Y.

24

I.

Jesus is Condemned to Death

Jesus took upon himself the guilt and anguish of all humanity,
so that sins may be forgiven.

Jesus Christ, is the only Savior of the human race,
for only those who will be reconciled to Him.
Jesus' road to Calvary is our road to Salvation,
for those who obey His will.

Human Trafficking, the enslavement and torture
of a fellow human being is a grave sin,
subject to eternal Hellfire.

Traffickers view their slaves as objects for
their own use and disposal.
Traffickers sell their slaves,
without regard or concern to the harm that will follow.

The slaves are sold for labor, organs, and sex.
Such slavery is the most heinous acts of sin known to man.

There is no recompense for the traffickers in these
gravest of all sins.

The traffickers will suffer eternal Hellfire for the
sins the commit against these enslaved victims.

Jesus and Mary, help the victims of trafficking.

II.

Jesus Carries His Cross

Jesus willingly takes up this Cross,
knowing what pain is to come for Him.

Jesus is willing to save souls,
if they repent of their sins and
turn from their ways.

What will it take for the traffickers and their
patrons, to see their sins,
and change their ways.

What will show these traffickers,
the harm they are doing to the
body and mind of their enslaved.

What will illumine theses traffickers,
to a life beyond their
self-pleasure and corruption.

What will resonate for these traffickers,
to teach them of a life beyond
this world,
in an Eternity to come.

Jesus and Mary, help the victims of trafficking.

III.

Jesus Falls the First Time

Do they know good from evil?
Do these Human Traffickers know
that their actions are evil?

All goodness, love, purity, and sacrifice,
comes to us from God.

But traffickers turn from God,
they deny their conscience,
they deny any goodness,
they live in dark evil.

Do you hear the call of the Lord?
Turn from your ways,
and behold the Lord.
The Lord gave His life that you may turn from sin.

Turn from your ways,
and save the lives of your victims.

Turn from your ways,
and ponder the suffering to
which you shall see in
an eternity of Hellfire.

Jesus and Mary, help the victims of trafficking.

IV.

Jesus Meets His Sorrowful Mother

Consider the families of these
missing men, women, and children.

Their families who grieve
in horror at what is done to their beloved.

These families are helpless,
to know where their loved ones
have gone,
and with what torture they face.

The Virgin Mary,
Mother of all the Faithful,
in sorrow for the Cross which her Son endures.

Virgin Mother,
most pure and obedient to the
law of the Lord,
sees the suffering of her Son on the Cross.

Our Jesus,
who offered himself,
Body and Blood,
Soul and Divinity,
for the conversion of sinners.

Jesus and Mary, help the victims of trafficking.

V.

Simon Helps Jesus to Carry the Cross

Praise be to God
for the law enforcement officials
who are searching to free
the slaves of Human Trafficking.

Praise be to God
for the Christians
who are educating the public to
free the slaves of Human Trafficking.

Praise be to God
for the professionals
who are vigilant
to the appearance of a poor slave
of Human Trafficking.

Come Holy Spirit,
fill the hearts of these Your helpers,
and enkindle in them the Fire of Your love.
Send forth Your Spirit,
and they shall uncover and renew
the life of these victims.

Jesus and Mary, help the victims of trafficking.

VI.

Veronica Wipes the Face of Jesus

See the imprint of the evil
and darkness
spreading our world,
as Human Traffickers
destroy these children of God.

If only we could wipe away the evil,
the sin of this trafficking,
the destruction,
the pain, suffering, and horror.

O God, make us ever more sensitive
to this evil in our world,
that we may be examples
only of the purity and morality
which You teach us.

O God, make us every more aware
to this evil in our world,
that we may report, intervene,
and put an end to this horror
for all time.

Jesus and Mary, help the victims of trafficking.

VII.

Jesus Falls the Second Time

They know good from evil.
These Human Traffickers must know
that their actions are evil.

All goodness, love, purity, and sacrifice,
comes to us from God.

But traffickers turn from God,
they deny their conscience,
they deny any goodness,
they live in dark evil.

Do you hear the call of the Lord?
Turn from your ways,
and behold the Lord.
The Lord gave His life that you may turn from sin.

Turn from your ways,
and save the lives of your victims.

Turn from your ways,
and ponder the suffering to
which you shall see in
an eternity of Hellfire.

Jesus and Mary, help the victims of trafficking.

VIII.

Women of Jerusalem Weep Over Jesus

We are all like the weeping women
of Jerusalem,
We are the weeping women
for the victims of Human Trafficking.

Unlike our Lord Jesus' suffering,
these victims are just victims.

Jesus calls us all to reach out to the suffering,
and to assist them in all ways.

Let us turn our tears to prayer,
turn our tears into education,
turn our tears into awareness,
turn our tears into saving lives,
turn our tears into laws and prosecution,
turn our tears into moral living.

Lord Jesus,
this Human Trafficking
crosses states, countries and borders.
They are on transportation across the world.
Let us feel the suffering of your Cross,
when we pass by
the enslaved and evil persecutors,
and know the action we should take to intervene.

Jesus and Mary, help the victims of trafficking.

IX.

Jesus Falls the Third Time

Again and again traffickers and their merchants
victimize the enslaved.
These human traffickers live
in grave evil.

All goodness, love, purity, and sacrifice,
comes to us from God.

But traffickers turn from God,
they deny their conscience,
they deny any goodness,
they live in dark evil.

Do you hear the call of the Lord?
Turn from your ways,
and behold the Lord.
The Lord gave His life that you may turn from sin.

Turn from your ways,
and save the lives of your victims.

Turn from your ways,
and ponder the suffering to
which you shall see in
an eternity of Hellfire.

Jesus and Mary, help the victims of trafficking.

X.

Jesus is Stripped of His Garments

Garments are the only protection
of our body that we have.

What makes us human is our conscience
and awareness that we are among
one another in the world.
Our garments protect our skin,
and they protect our morality.

But victim of
Human Trafficking
are wearing scant garments,
often of dirty, disheveled, and immoral regard.

The victims are prey
to their traffickers and merchants.
The victims endure
poverty, torture, abuse, rape.

Just as Jesus had no
shield of protection,
so too the enslaved have no protection.

Mother, most pure,
protect God's children from such harm.

Jesus and Mary, help the victims of trafficking.

XI.

Jesus is Nailed to the Cross

Hear the pounding of the hammer,
and the crushing of Jesus' bones.

But Jesus still lives,
He lives in all of us,
those of whom follow the
will of our God.

Jesus does not deny his role
as our Savior.
Jesus accepts His suffering on the Cross
for the salvation of His people.

Jesus endures
as the nails are hammered
through his flesh and bones.

Hear the pounding of the hammer,
just as the pounding hearts and
tortured bodies of the victims of trafficking.

The enslaved are beating along with hammer
on Jesus' cross.
Feel the unimaginable pain
that does not stop,
that continues on
every day in these victims.

Jesus and Mary, help the victims of trafficking.

XII.

Jesus is Crucified

Ponder the five wounds
of Jesus Christ
as He is taken up on the Cross.

Jesus looked down from the Cross,
just as He now looks down from Heaven.

Jesus took upon Himself
the agony of the Cross,
so that goodness
would flourish in the world.

Yet, there is only evil in Human Trafficking.

See Jesus holy head dripping
in blood, sweat and tears,
as the thorns pierce Him.
See Jesus holy hands and feet
impaled with metal nails
that have torn through His being.
See Jesus side
pierced with a sword,
at which flow blood and water
for the life and salvation of His people.

Jesus took upon Himself
the sins of the world,
for those that will follow His will.

Traffickers turn from your ways.

Jesus and Mary, help the victims of trafficking.

XIII.

Jesus is Taken Down from the Cross

The skies thundered
and the ground shook,
"Surely this was the Son of God."

The earth and heavens thunder
and tremble as
the evil of Human Trafficking goes on.

Traffickers, do you not see
that God and all goodness
reign above?

Traffickers, can you imagine
what Hell you will endure
for all eternity,
if you do not turn from your ways at once.

There is no salvation for
the evil among us.
The thunderous fear will endure
for all eternity for the evil among us.

Jesus, only Son of God, save your people.

Jesus and Mary, help the victims of trafficking.

XIV.

Jesus is Laid in the Tomb

Jesus is dead,
His Mother and disciples
grieve in great sadness
as they wrap his holy Body and
lay it in the tomb.

But this great sadness
will be no more,
because, as it was promised:
on the third day,
Jesus rose from
the grave.

Jesus Christ is risen!
Jesus rose from the dead,
and is seated at the right hand of the Father.

From that time on,
Jesus reigns to judge the living and the dead.

Traffickers, do you not consider,
that Jesus sees your actions
at all times and in all places?

Traffickers, you will have no place to hide,
when the day of judgement comes for you.

Jesus and Mary, help the victims of trafficking.

Litany to the Holy Spirit for Human Trafficking

Lord, have mercy on them.
Christ, have mercy on them.
Lord, have mercy on them.

Father all-powerful, have mercy on them. Jesus, Eternal Son of the Father, Redeemer of the world, save them. Spirit of the Father and the Son, boundless life of both, sanctify them. Holy Trinity, *hear us*.

Holy Spirit, Who proceeds from the Father and the Son, enter their hearts. Holy Spirit, Who art equal to the Father and the Son, enter their hearts.

Promise of God the Father, *Have mercy on them.*
Ray of heavenly light, *Have mercy on them.*
Author of all good, *Have mercy on them.*
Source of heavenly water, consuming fire. *Have mercy on them.*
Ardent charity. *Have mercy on them.*
Spiritual unction. *Have mercy on them.*
Spirit of love and truth. *Have mercy on them.*
Spirit of wisdom and understanding. *Have mercy on them.*
Spirit of counsel and fortitude. *Have mercy on them.*
Spirit of knowledge and piety. *Have mercy on them.*
Spirit of the fear of the Lord. *Have mercy on them.*
Spirit of grace and prayer. *Have mercy on them.*
Spirit of peace and meekness. *Have mercy on them.*
Spirit of modesty and innocence. *Have mercy on them.*
Holy Spirit, the Comforter. *Have mercy on them.*
Holy Spirit, the Sanctifier. *Have mercy on them.*
Holy Spirit, Who governs the Church. *Have mercy on them.*
Gift of God, the Most High. *Have mercy on them.*
Spirit Who fills the universe. *Have mercy on them.*
Spirit of the adopted children of God. *Have mercy on them.*

Holy Spirit, inspire them with horror of sin. *Amen.*
Holy Spirit, come and renew the face of the earth. *Amen.*
Holy Spirit, shed Thy light in their souls. *Amen.*
Holy Spirit, engrave Thy law in their hearts. *Amen.*
Holy Spirit, inflame them with the flame of Thy love. *Amen.*
Holy Spirit, open to them the treasures of Thy graces.
Amen.
Holy Spirit, teach them to pray well. *Amen.*
Holy Spirit, enlighten them with Thy heavenly inspirations.
Amen.
Holy Spirit, lead them in the way of salvation. *Amen.*
Holy Spirit, grant them the only necessary knowledge.
Amen.
Holy Spirit, inspire in them the practice of good. *Amen.*
Holy Spirit, grant them the merits of all virtues. *Amen.*
Holy Spirit, make them persevere in justice. *Amen.*
Holy Spirit, be Thou their everlasting reward. Amen.

Lamb of God, Who takes away the sins of the world.
Send them Thy Holy Spirit.
Lamb of God, Who takes away the sins of the world, *Amen.*
pour down into their souls the gifts of the Holy Spirit. *Amen.*
Lamb of God, Who takes away the sins of the world.
grant them the Spirit of wisdom and piety. *Amen.*

L. Come, Holy Spirit! Fill the hearts of Thy faithful,
R. *And enkindle in them the fire of Thy love.*

Let us pray. Grant, O merciful Father, that Thy Divine Spirit
may enlighten, inflame and purify them, that He may
penetrate them with His heavenly dew and make them
fruitful in good works, through Our Lord Jesus Christ, Thy
Son, Who with Thee, in the unity of the same Spirit, lives and
reigns, one God, forever and ever.

R. *Amen.*

Identifying

Human Trafficking

Victims

NOT FOR SALE

Who is at risk for being trafficked?

- children and youth travelling alone
- vulnerable or runaway teenagers
- those seeking a "better life"
- the lonely, seeking a relationship
- those with sexual identity confusion
- those in abusive relationships
- immigrants and language barriers
- those seeking to immigrate
- impoverished families who are misled
- drug users and wandering homeless

Where might anyone see trafficking victims?

- hospital emergency rooms
- urgent care clinics
- free health clinics
- quick shop retail stores
- unusual spa parlors
- bars, clubs, venues
- buses, planes, toll roads, taxis, ships
- street corners, run down housing
- state and national borders
- hotels and motels

What can I do for them?

- Call 911 if possible without creating a problem
- Contact the anonymous, confidential hotline
  1 (888) 373-7888
  text **HELP** or **INFO** or **BEFREE**
  24 hours a day/7 days per week
  200 language interpreters available
- provide any detail you can when calling, which may include skin color, clothing, tattoos, vehicle

humantraffickinghotline.org

Safety planning can include "risk assessments, preparations, and contingency plans to increase the safety of a human trafficking victim or an individual at-risk for human trafficking, as well as any agency or individual assisting a victim."

Safety plans:

1. Assess the current risk and identify current and potential safety concerns

2. Create strategies for avoiding or reducing the threat of harm

3. Outline concrete options for responding when safety is threatened or compromised

(humantraffickinghotline.org)

What are the indicators someone may be a victim?

- no identification
- no health insurance
- domineering person overseeing care
- poorly or scantily clothed
- not knowing where they are or come from
- family background is unknown
- delay in answering basic questions
- fearful, sickly, and injured

What physical symptoms will victims have?

- malnourished, dehydrated
- scars, bruises, burns, wounds
- 'ownership' markings like a number or symbol
- sexually transmitted disease
- HIV/AIDS
- genital tears, scars, inflammation
- pregnancy
- pain, strain, weakness
- vague reference to pain
- rashes

What mental symptoms will victims have?

- fatigue
- illiteracy
- fearful, withdrawn, confused
- misplaced ideas of who is good/bad
- hysteria
- anger
- depression or suicidality
- panic
- cultural shock
- hallucinations
- PTSD

Call 1 (888) 373-7888

Freedom Network USA sites that:

First responders who are particularly well-positioned to identify and support trafficking victims include: criminal law enforcement agencies that work with sexual assault and domestic violence survivors; child advocacy centers; sexual assault service providers, sexual assault nurse examiners; health care providers; and child and social welfare agencies; and legal/social services agencies that deal directly with labor violations, which have the ability to screen for trafficking and assist with U-visas and T-visas and other forms of legal relief. *(freedomnetworkusa.org)*

Laws

to end

Human Trafficking

Pub. L. 115-398
Dec. 31, 2018
132 Stat. 5328

H. R. 767

One Hundred Fifteenth Congress
of the
United States of America

AT THE SECOND SESSION

*Begun and held at the City of Washington on Wednesday,
the third day of January, two thousand and eighteen*

An Act

To establish the Stop, Observe, Ask, and Respond to Health and Wellness Training
pilot program to address human trafficking in the health care system.

*Be it enacted by the Senate and House of Representatives of
the United States of America in Congress assembled,*

SECTION 1. SHORT TITLE.

This Act may be cited as the "Stop, Observe, Ask, and Respond
to Health and Wellness Act of 2018" or the "SOAR to Health
and Wellness Act of 2018".

SEC. 2. PROGRAM ESTABLISHMENT.

Part E of title XII of the Public Health Service Act (42 U.S.C.
300d–51 et seq.) is amended by adding at the end the following:

"**SEC. 1254. STOP, OBSERVE, ASK, AND RESPOND TO HEALTH AND
WELLNESS TRAINING PROGRAM.**

"(a) IN GENERAL.—The Secretary shall establish a program
to be known as the Stop, Observe, Ask, and Respond to Health
and Wellness Training Program or the SOAR to Health and
Wellness Training Program (in this section referred to as the 'Program') to provide training to health care and social service providers
on human trafficking in accordance with this section.

www.uscode.house.gov for full bill summary

Stop, Observe, Ask, and Respond to Health and Wellness Act of 2017

(SOAR) to Health and Wellness Act of 2017

(Sec. 3) This bill directs the Department of Health and Human Services (HHS) "to establish a program, to be known as the Stop, Observe, Ask, and Respond to Health and Wellness Training Program or the SOAR to Health and Wellness Training Program", to train health care providers and other related providers to:

1. identify potential human trafficking victims

2. work with law enforcement to report and facilitate communication with such victims

3. refer victims to social or victims service agencies or organizations

4. provide such victims with coordinated care tailored to their circumstances

5. consider integrating this training with existing training programs

The program "must include the functions of the training program with the same name that was operating before this bill's enactment and the following initiatives":

*   engaging stakeholders to develop a flexible training module

*   providing technical assistance to health education programs and health care professional organizations

*   facilitating the dissemination of best practices

*   developing a methodology for collecting and reporting data on the number of human trafficking victims served in health care settings or other related provider settings

www.uscode.house.gov for full bill summary

<u>SOAR</u> Training Teaches

- **Stop** – Describe the scope of human trafficking in the United States
- **Observe** – Recognize the verbal and non-verbal indicators of human trafficking
- **Ask** – Identify and interact with individuals who have experienced trafficking using a victim-centered and trauma-informed approach
- **Respond** – Respond effectively to potential human trafficking in your community by identifying needs and available resources to provide critical support and assistance

Who should take the SOAR training?

- Health care providers
- Social workers
- Public health professionals
- Behavioral health professionals

Office of Trafficking in Persons, U.S. Department HHS,
An Office of the Administration of Children and Families
www.acf.hhs.gov/otip/training/soar-to-health-and-wellness-training

➤ Professional Development Scholarships
➤ Organizational Scholarship
➤ Human Trafficking Leadership Academy
➤ *Look Beneath the Surface* Regional Programs
➤ Free Resource Materials

- Immigration and Customs Enforcement:
  Human Trafficking and Human Smuggling

  https://www.ice.gov/features/human-trafficking
  Report Crimes:
  Email https://www.ice.gov/webform/hsi-tip-form or
  Call   1-866-DHS-2-ICE

- DHS: Blue Campaign/Human Trafficking

  https://www.dhs.gov/blue-campaign

- DHS: U Visa Law Enforcement Resource Guide

  https://www.dhs.gov/publication/u-visa-law-
  enforcement-certification-resource-guide

- Customs and Border Protection: Human
  Trafficking

  https://www.cbp.gov/border-security/human-
  trafficking

- Federal Law Enforcement Training Centers:
  Human Trafficking Training Program
  https://www.fletc.gov/human-trafficking-
  training-program

- Polaris Project

  Polaris is named for the North Star, which
  people held in slavery in the United States used
  as a guide to navigate their way toward
  freedom.
  www.polarisproject.org/action (Take Action)
  https://polarisproject.org/current-federal-laws

- National Crime Victims' Rights Week
  April, sponsored by OVC

Polaris' list of Federal Laws to End Trafficking

-National Defense Authorization Act of 2013
Sections 1701-1708 of the National Defense Authorization Act seeks to end human trafficking that could be associated with government contractors.

-Trafficking Victims Protection Reauthorization Act (TVPRA) of 2000, 2003, 2005, 2008, 2013
This law, in continuous amendment and improvement, is the cornerstone of Federal human trafficking legislation, and established several methods of prosecuting traffickers, preventing human trafficking, and protecting victims and survivors of trafficking.

-The PROTECT Act
The Prosecutorial Remedies and Other Tools to End the Exploitation of Children Today (PROTECT) Act of 2003, established enhanced penalties for individuals engaging in sex tourism with children, both within the United States and in other countries.

-The Mann Act of 1910, (18 U.S.C. § 2421-2424) as amended in 1978 and again in 1986, criminalizes the transportation of minors, and the coercion of adults to travel across state lines or to foreign countries, for the purposes of engaging in commercial sex.

-The Tariff Act of 1930
The Tariff Act of 1930 prohibits importing goods made with forced or indentured labor.

-The Racketeering Influenced Corrupt Organizations Act (RICO)
RICO was created to be a tool for the federal government to more effectively prosecute members of organized crime for racketeering offenses.

-The Customs and Facilitations and Trade Enforcement Act (2009)
This act amended the prohibition on importing goods made with slave or indentured labor to include goods made through the use of coercion or goods made by victims of human trafficking.

Interventions

for

Human Trafficking

National Referral Directory Application

To add your service agency to a provider list for victims of human trafficking, you must complete an application form at:

https://forms.humantraffickinghotline.org/4

email:

help@humantraffickinghotline.org

Please be aware that most people can only intervene by means of calling the Human Trafficking Hotline number or 911. No one should put themselves in danger to intervene, and are likely not trained to physically intervene.

If the appropriate authorities have intervened and separated the victim from the traffickers, there are steps to help the victims, depending on where you work or volunteer.

If you are working or volunteering in a capacity where you are providing care to a victim, you must be very sensitive to their crisis situation.

- ✓ Speak gently
- ✓ Remain an appropriate distance from them
- ✓ Slowly explain that you are there to help them
- ✓ Provide direction on the services to ensue

Victims will likely need the following care, in order of urgency:

- ➢ Language interpreter
- ➢ Safety plan/person of professional authority
- ➢ Health/Medical Care/Rx
- ➢ Urology/Gynecology/Gastroenterology
- ➢ Wound Care
- ➢ Mental Health Care/Psychiatry/Neurology
- ➢ Dental, Ophthalmology, Podiatric Care
- ➢ Counseling
- ➢ Victims Services/Legal
- ➢ Housing and Shelter
- ➢ Clothing, Underwear, Socks, Shoes
- ➢ Toiletries
- ➢ Notebook/Pen/Markers
- ➢ Hair Care
- ➢ Transportation
- ➢ Food

EMTALA

In 1986, the Center for Medicare and Medicaid Service, and Congress, enacted a law called EMTALA. EMTALA stands for <u>Emergency Medical Treatment & Labor Act</u>.

Under EMTALA law, hospitals are required to provide stabilizing treatment for patients with emergency medical conditions, regardless of their ability to pay the hospital. If a hospital is unable to medically stabilize a patient within its capabilities, or if the patient requests, an appropriate transfer to a different hospital should be implemented.

The U.S. government, at the federal and state levels, make possible reception of services to victims of human trafficking, regardless of their situation.

The following services are available to **all victims**, according to general eligibility guidelines.

- Child Nutrition Program (USDA)
  https://www.fns.usda.gov/school-meals/child-nutrition-programs
- (WIC) Supplemental Nutrition for Women and Infants
  https://www.fns.usda.gov/wic/women-infants-and-children-wic
- Health Resources and Services Administration
  https://www.hrsa.gov/get-health-care/index.html
- Substance Abuse and Mental Health Services Administration
  https://www.samhsa.gov/find-treatment
- Victims of Crime Act (VOCA) Victim Compensations
  https://www.acf.hhs.gov/otip/victim-assistance/services-available-to-victims-of-trafficking
- Career One Stop (training and jobs)
  http://www.careeronestop.org/
- Federal Victim–Witness Coordination
  https://www.justice.gov/usao/find-your-united-states-attorney
- Emergency Witness Assistance
- Witness Security Program
  http://www.usmarshals.gov/witsec/
- Survivors of Torture
  https://www.acf.hhs.gov/orr/programs/survivors-of-torture
- Office for Victims of Crimes (OVC) Human Trafficking
  https://ovc.ncjrs.gov/humantrafficking/traffickingmatrix.html

This public information is obtained from the government websites as indicated.

Other services offered by the government, are broken down accordingly by age and citizenship status.

1. U.S. Citizens

Who is a citizen?

An "individual born in the United States, Puerto Rico, Guam, the Commonwealth of the Northern Mariana Islands, the U.S. Virgin Islands, American Samoa, or Swain's Island"; **or** "foreign-born children <u>under age 18</u> residing in the U.S. with their birth or adoptive <u>parents</u>, at least <u>one of whom is a U.S. citizen</u> by birth or naturalization; or individuals granted citizenship status by the Immigration and Naturalization Services (INS)." *(www.acf.hhs.gov)*

(SNAP) Supplemental Nutrition Program/
Food Stamps (County Assistance-Welfare Offices)
https://www.fns.usda.gov/snap/supplemental-nutrition-assistance-program-snap
Public Housing (HUD)/Tenant-Based Vouchers
https://www.hud.gov/topics/rental_assistance/phprog
Medicaid (by state)
https://www.medicaid.gov/about-us/contact-us/contact-state-page.html
Children's Health Insurance Program (CHIP)
https://www.medicaid.gov/chip/index.html
Temporary Assistance for Needy Families (TANF)
https://www.acf.hhs.gov/ofa/help
Supplemental Security Income (SSI)
https://www.ssa.gov/ssi/
Job Corps
https://www.jobcorps.gov/
Title IV Federal Student Financial Aid
https://studentaid.ed.gov/sa/

2. Individuals Lawfully Present in the U.S.,

but not U.S. Citizen or Lawful Permanent Resident

"An individual paroled for at least one year who the government has agreed not to remove from the United States for a temporary period. This includes nonimmigrants who are admitted to the United States on a temporary basis, such as a person on a student visa, exchange visitor visa, or temporary worker visa." *(www.acf.hhs.gov)*

- Medicaid
- (CHIP) Children's Health Insurance Program

3. Refugee, Asylee, or Cuban/Haitian Entrant

"An individual granted refugee or asylee status by the U.S. Refugee Admissions Program (USRAP), or, an individual granted parole status as a Cuban/Haitian entrant under the Cuban Haitian Entrant Program (CHEP)."

Every state has a **State Refugee Coordinator or Regional Representative** who oversees benefits for trafficking victims, as well as refugees and other populations.

Similarly to U.S. Citizens, they are able to get:

- Supplemental Nutrition Assistance Program/Food Stamps (SNAP) (as above web address)
- Public Housing Program (as above web address)
- Tenant-Based Vouchers (as above web address)
- Children's Health Insurance Program (as above web address)
- Refugee Medical Assistance *https://www.acf.hhs.gov/orr/programs/cma/about*
- Office of Refugee Resettlement (ORR) Medical Screenings *https://www.acf.hhs.gov/orr/programs/preventive-health/about*

- Temporary Assistance for Needy Families (TANF)
  (as above web address)
- Refugee Cash Assistance
  https://www.acf.hhs.gov/orr/programs/cma/about
- Supplemental Security Income (SSI)
  (as above web address)
- Job Corps (as above web address)
- Title IV Federal Student Financial Aid
  (as above web address)
- Refugee Social Services and Targeted Assistance
  https://www.acf.hhs.gov/orr/programs/refugee-social-services/about
- Voluntary Agency Matching Grant Program
  https://www.acf.hhs.gov/orr/programs/matching-grants/about
- Unaccompanied Refugee Minors (URM) Program
  https://www.acf.hhs.gov/orr/resource/unaccompanied-refugee-minors

4. Adult HHS Certification Letter
   https://www.acf.hhs.gov/otip/victim-assistance/certification-and-eligibility-letters-for-foreign-national-victims

"Foreign national adults now in the United States who have been subjected to trafficking in persons are eligible for certain benefits and services under the Trafficking Victims Protection Act. DHHS Certification Letters allow individuals who have experienced trafficking, and who meet certain eligibility rules, apply for the same benefits and services as refugees." *(www.acf.hhs.gov)*

1-866-401-5510

5. Minors with a DHHS Interim Assistance or Eligibility Letter
https://www.acf.hhs.gov/otip/victim-assistance/eligibility-letters

"Foreign national minors now in the United States who have been subject to trafficking in persons are eligible for certain benefits and services under the <u>Trafficking Victims Protection Act</u>. DHHS Interim Assistance and Eligibility Letters allow minors who have experienced human trafficking, and meet certain eligibility rules, apply for the same <u>benefits and services</u> as refugees." *(www.acf.hhs.gov)*

- **Trafficking Victim Assistance Program** grantees can provide case management services to assist foreign nationals with HHS certification and enrollment in benefits.
  1-866-401-5510
- **ORR State Refugee Coordinators** can help navigate state-specific benefits questions.
- **National Human Trafficking Hotline** is available 24 hours a day, 7 days a week for technical assistance and service referrals. Call 1-888-373-7888 or email **help@humantraffickinghotline.org**.

humantraffickinghotline.org

As stated, National Human Trafficking Hotline offers a list of registered community service programs for referral needs.

If you are an organization that wants to become a registered service provider, follow the directions on **page 52** of this book.

For all other needed community services, consider the following tips.

- ➤ All minors under age 18, and in some cases, under age 21, should be immediately referred to your County children and youth services (CYS) intake. They will assist with emergency intake and fast home placement, if a hospital stay is not required.
- ➤ Many counties have independent youth foster care placement programs that work with the county children and youth programs. They can place the minor in a home the same day, if a hospital stay is not required.
- ➤ There are many residential and group homes for troubled youth, in which CYS may be able to find emergency youth shelter placement.
- ➤ There are confidential hidden shelters in most counties for adults who are facing violence.
- ➤ Health Resources and Services Administration *https://www.hrsa.gov/get-health-care/index.html* These listed health care institutions and programs are usually available free of charge to victims of trafficking who do need health interventions.
- ➤ Victims of trafficking may need to ultimately get back to where they came from, whether it be to a family, different state, or different nation. Adults have a choice to return or not. Children will be encouraged to return, if safe to do so, and CYS can work on family reunification.
- ➤ If an adult chooses to stay in the location they are freed in, and they are not a legal citizen, they will have to follow the standard process to gain citizenship. There is risk of relapse[1] in these cases.

1. See Recovery and Reintegration, page 72

(PTSD)

Post-Traumatic Stress Disorder

What is a qualitative assessment?

Establish rapport

Non-Judgmental Attitude

Read Non-Verbal

Believe Disclosure

Trauma-Informed and Sensitive

Professional Competence

T.Riebel, 2016

What is a Trauma-Informed approach?

1. Reduce re-traumatization
2. Highlight survivor strengths and resilience
3. Promote healing and recovery
4. Support the development of good health, and short and long-term coping mechanisms.

Seeking safety for the patient includes professional knowledge of four content areas:

➢ Cognitive
➢ Behavioral
➢ Interpersonal
➢ Case Management
➢ Attention to clinical processes

However, safety can also be expressed in compassion and genuine concern for the patient-victim.

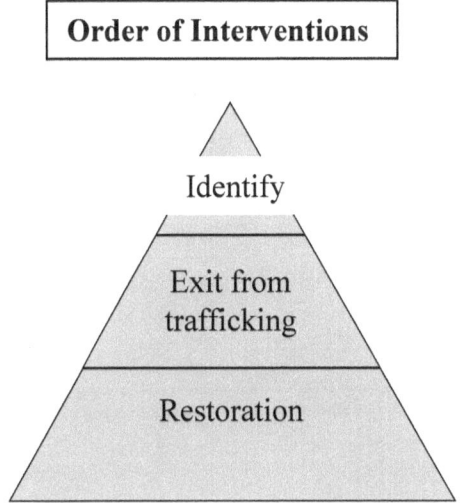

65

Health Care Workers

Primary Immediate Goals

❖ Be able to identify, report, treat, suspected victims of trafficking

❖ Notify your Manager,
   Social Work Department,
   Security Officers

❖ Identify your patient's ICD-10 Diagnostic Codes

❖ Maintain professional continuing education programs respective to your field of practice

What is (PTSD) Post-Traumatic Stress Disorder?

There are two commonly used diagnostic manuals of codes, **ICD-10** which is used in medical settings, and the **DSM-V** which is used in psychiatric or mental health settings.

According to the ICD-10 (2015), PTSD can be coded in three different ways:

- F43.10, post-traumatic stress disorder, unspecified

- F43.11, post-traumatic stress disorder, acute

- F43.12, post-traumatic stress disorder, chronic

According to the DSM-V (2013), PTSD for children (over age 6), adolescents and adults is coded as:

- 309.81
  Posttraumatic Stress Disorder
  (The criteria listed in the DSM-V book are extensive, and as such, mental health practitioners should consult it specifically)

Further, specifications as to whether PTSD with:

- Dissociative symptoms: The individual's symptoms meet the criteria for posttraumatic stress disorder, and in addition, in response to the stressor, the individual experiences persistent or recurrent symptoms of either of the following:

  1. Depersonalization: Persistent or recurrent experiences of feeling detached from, and as if one were an outside observer of, one's mental processes or body (e.g., feeling as though one were in a dream; feeling a sense of unreality of self or body or of time moving slowly).

2. Derealization: Persistent or recurrent experiences of unreality of surroundings (e.g., the world around the individual is experienced as unreal, dreamlike, distant, or distorted).

*Note: To use this subtype, the dissociative symptoms must not be attributable to the physiological effects of a substance (e.g., blackouts, behavior during alcohol intoxication) or another medical condition (e.g., complex partial seizures). *DSM-V, 2013*

And further specification as to:

• With delayed expression: If the full diagnostic criteria are not met until at least 6 months after the event (although the onset and expression of some symptoms may be immediate). *DSM-V, 2013*

Any victim of human trafficking will get a PTSD diagnosis, because their trauma has been so severe. The PTSD symptoms will likely remain with the victim for the remainder of their lives, to varying degrees.

Psychotherapy may help with the management and ease of Post-Traumatic Stress Disorder symptoms. Psychiatric medication, prescribed by a physician, will also help ease the severity of some of these traumatic feelings. Victims should seek treatment from the most qualified providers, who are licensed and specialists in the field of trauma.

Any person who is diagnosed with PTSD can also have another mental or medical condition. These are called co-morbid or co-occurring disorders. An example of co-morbid disorders could be: PTSD and Major Depression.

Some symptoms exhibited from PTSD can also be due to a medical condition, or even just cultural shock. (Providers may consider differential diagnosis.) Such other conditions may cause and mimic dissociation or derealization. But certainly with the severity of PTSD, as a result of being victim of human trafficking, dissociation and derealization are highly common and likely.

The Department of Health and Human Services (DDHS), sites,

> In addition to experiencing terrorizing physical and sexual violence, researchers report that victims often experience multiple layers of trauma including psychological damage from captivity and fear of reprisals if escaped,, brainwashing, and for some, a long history of family, community, or national violence (Stark & Hodgson, 2003; Ugarte, Zarate, & Farley, 2003).

DHHS states, "this impact of trauma can make the job of first responders and those trying to help victims difficult at best."

> Post-trauma responses including difficulties controlling emotions, sudden outbursts of anger or self-mutilation (Briere & Gil, 1998), difficulties concentrating, suicidal behaviors (Zlotnick, Donaldson, Spirito, & Pearlstein, 1997), alterations in consciousness (dissociation), and increased risk-taking. These post-traumatic symptoms and problems are those which service providers identify as common among the trafficking victims they encounter. For some victims, in particular victims of sex trafficking, the use of alcohol and drugs to escape these emotional states is also a problem.

The characterizing symptoms of PTSD include, but are not limited to: intrusive re-experiencing of the trauma (e.g., flashbacks, nightmares, disturbing thoughts), numbing of traumatic stimuli (e.g. avoiding places, people, and situations), and hyper arousal (e.g., heightened startle response, and difficulty concentrating).

Las Vegas Review-Journal reported that Las Vegas Fire Department firefighter-paramedic M. Driscoll used his extra time to create a program that would help responders, as well as social workers and hospital employees, to recognize the signs of trafficking victims.

iEmpathize, an organization that works to eradicate sex trafficking. **Mission**: We equip adults to empower youth to eradicate exploitation. Learn about the child-centered approach. *iempathize.org*

Recovery

and

Reintegration

United Nations
Trust Fund

<u>Victim-Centered</u>

Prevention
Prosecution
Protection

NGO services are funded in
30 countries globally
through grants.

<u>www.unodc.org</u>

UNODC
United Nations Office on Drugs and Crime

Recovery

Recovery for the victims of human trafficking is a process. Depending on the severity of their trauma, other health complications, pre-existing conditions, and personal coping abilities, all effect the recovery process of the victims.

The victims of human trafficking have earned the right to a sensitive and concerned advocacy in their road to recovery.

Medical emergencies, psychiatric crisis, and shelter may take immediate precedence. If the victim needs to be hospitalized for a few days, and then transferred to a step down level of residential care, care managers will have a few extra days to secure shelter and connect with the necessary systems or family members.

- Refer to page 60 of this book on seeking interventions for children, adolescents, and domestic violence shelters.
- For nursing homes and specialty care centers, refer to the link on page 55 Health Resources and Service Administration.
- All of the federal benefit options are in the Interventions section of this book beginning on page 51.

Victims of human trafficking will not all present in the same way, nor will they recover in the same way.

Additionally, common therapy techniques for other types of trauma victims may not be suitable for human trafficking trauma. It is always important to proceed with sensitivity and at the victim's direction.

Safety and stabilized health is the primary short and long-term goal for these permanently traumatized people. If they wish to remain in a passive but safe and stabilized capacity, then let them. But if the victims do want to pursue age-appropriate life activities, then you may begin to counsel them accordingly, using best practices. Some victims have gone on to obtain college degrees, speak, and write books of their story.

Reintegration
Development of a Home Visitor/Advocate/Case Manager Program

We are lucky in the United States to have social service and community support agencies; not all countries provide these human services. Human service agencies in the community have the important role to develop specialty programs for victim care. The most effective program will consist of an ongoing worker(s) to be involved with the victim for probably the remained of their life.

Such an ongoing worker may be a nurse, social worker, victim advocate, or case manager. These roles will frequently interact with the victim to make sure that they are doing alright and staying safe. The workers duties may consist of the following:
1. Medication management
2. Doctor/Provider follow-through
3. Insurance and benefit coverage
4. Family or residential counseling or visits
5. Treatment planning
6. Housing transitions
7. Being a trustworthy safe person

Workers taking on care for these victims, should ideally be with the victim as long as they can, versus frequent worker changes, which would disrupt feelings of security and life stability. Workers should be knowledgeable about trauma-informed care, so that there is no risk of re-traumatization. This would most certainly involve as a basic, a reliable, gentle, consistent, caring, and pleasant professional, aware of the boundaries necessary for such a traumatized person.

The risk of *relapse* is high for any victim who is not stabilized in any of their major life categories, for example: housing, family/systems of care, food/shelter, drug or alcohol use, or abusive relationship situations, to name a few. This is why it is vital to have the support of an ongoing worker to oversee and manage life for these victims.

Executive Director and Co-Founder of (HEAL) Health, Education, Advocacy, Linkage, Hanni Stoklosa MD PMH is an advocate, researcher, and speaker on the wellbeing of trafficking survivors in the U.S. and internationally through a public health lens. She has advised the United Nations, International Organization for Migration, U.S. Department of Health and Human Services, U.S. Department of Labor, U.S. Department of State, and the National Academy of Medicine on issues of human trafficking. *(Harvard Catalyst/Harvard Humanitarian Initiative)*

Dr. Stoklosa presented a seminar at the Radcliffe Institute for Advanced study *Operational Guidance to Overcome Trafficking among Urban Migrants during Humanitarian Crises*. She says, that as a result of conflict, poverty, and other complex social, cultural, and political factors, the world today is facing greater rates of urban displacement and larger numbers of urban migrants than ever before.

In a PubMed listed journal, Reconstructing a Sense of Self: Trauma and Coping Among Returned Woman Survivors of Human Trafficking in Vietnam, author P.D. Le, researches the process of acculturation for such victims.

Le explains that the victims
> are "seeking congruence between their self-understandings and the changing contextual factors [since victimized] while exhibiting three main coping strategies: regulating emotional expression and thought, creating opportunities within constraints, and relating to cultural schemas. The findings underscore the importance of considering contextual factors such as cultural norms and societal values in efforts to assist trafficked survivors reintegrate into their communities (Le, 2017)."

This journal article is the just the tip of the surface in the immense and deep recovery which surviving victims will face. The culture for a victim of trafficking may refer to national or political culture, but also to family and community culture, culture of age range and time, and culture of being 'different'.

The government offers alternate options for some victims of human trafficking, who have the strength and ability to serve for the cause in which they were once entrapped. U.S. Citizenship and Immigration Services Program explains:

✓ <u>T Nonimmigrant Status (T-Visa)</u>
T nonimmigrant status provides immigration protection to victims of trafficking. The T Visa allows victims to remain in the United States and assist law enforcement authorities in the investigation or prosecution of human trafficking cases.

✓ <u>U Nonimmigrant Status (U-Visa)</u>
U nonimmigrant status provides immigration protection to crime victims who have suffered substantial mental or physical abuse as a result of the crime. The U visa allows victims to remain in the United States and assist law enforcement authorities in the investigation or prosecution of the criminal activity.
(www.uscis.gov)

Department of Homeland Security   *(www.dhs.gov)*
https://www.dhs.gov/sites/default/files/publications/U-Visa-Immigration-Relief-for-Victims-of-Certain-Crimes.pdf

An example of such a victim-advocate is Dr. Katariina Rosenblatt, LL.M, PhD, a U.S. citizen by birth, who endured trafficking in a hotel, her middle school and organized crime from the ages of 13-17. She found her hope in God and the strength to escape her suffering, and contribute to the cause. Dr. Rosenblatt, is the author of two books: Stolen (2014), and Trafficking In America (2014) and the founder of *There is HOPE for Me, Inc.* She is a Consultant and National Trainer for the DOJ and of the Train the Trainer & curriculum development for the Dept. Of Health and Human Services. She has also been a long-time member and speaker for the National Survivor's Network and Speaker's Bureau. She has spoken on the topic to both state and federal legislative agencies on Capitol Hill, the DOJ, State Department, HHS and HUD and speaks on trafficking and domestic violence to universities, law enforcement, public agencies such as departments of children and families nationwide, juvenile justice centers, and travel and tourism offices. *www.DrKat.net*

Faith Resources

for

Human Trafficking

JESUS THE TRUE SHEPHERD

Publ. & Print. by Th.Kelly, 17 Barclay St. N.Y.

EWTN radio talk show host of *The Good Fight,* recently said, "they are crushed beneath the weight of another's sins." That statement well describes the situations for the victims of human trafficking. While we have all sinned in our lives, there is nothing that these victims have done that would have caused them to be sold in trafficking. **The sin is in the buyers and users.**

Child of God, by Music for the Soul, writes:

> A song to help those who've been victims of sex trafficking or caught up in the sex trade to understand that there is nothing they have ever done – or anything that has ever been done to them – that can leave them outside of the love of God. *(musicforthesoul.org)*
>
> [Available in multiple languages]

Counsel victims to live in the presence, comfort, and healing grace of God. The God and Father of all creation cries out for the victims of trafficking. Seek the compassion of our Lord, in the gentle arms of His Blessed Mother, Mary. God cares for you, and wishes for you to be healed. There is no shame, only mercy.

St. Michael the Archangel,
defend them in battle.

Songs for the Victims of Human Trafficking

Child of God, Music for the Soul
Run, Chris Rogers & Sara Cariveau
She, Natmozzie

Be their protection against the wickedness and snares of the devil. May God rebuke them, we humbly pray; and do thou, O Prince of the Heavenly host, by the power of God, cast into hell, Satan and all the evil spirits, who prowl about the world seeking the ruin of souls. Amen

(USCCB) United States Conference of Catholic Bishops'
Become a Shepherd Toolkit

Stop
Human Trafficking and
Exploitation
Protect
Help
Empower and
Restore
Dignity

BECOME*a***SHEPHERD**

(usccb.org)

In a 2014 assembly with the Pontifical Academy of Sciences, Pope Francis said, "Human trafficking is an open wound on the body of contemporary society, a scourge upon the body of Christ, a crime against humanity."

Pope Francis, along with international leaders, religious, and victims gathered together to share stories and insights about the problem of human trafficking. They urged that the church and police [law] need to work together to fight this problem.

Later in a 2016 meeting at the Vatican, Pope Francis said, "An unrelenting, coordinated commitment is needed to prevent people from falling prey to traffickers and to help victims caught in their snares." He continued, "[victims] are stripped of their dignity, physical and mental integrity and sometimes even their life," Pope Francis reiterated the horror of such evils.

The Catholic Church leaders in the United States, or USCCB, offer a toolkit to individuals or groups to begin education and awareness for the fight to end human trafficking. Email: **MRSShepherd@usccb.org** to request the online SHEPHERD toolkit, sample presentation, and leader's guide to help get started with your parish or diocesan group.
(usccb.org)

There is a significant amount of related information on trafficking, immigration, faith petitions, and world mission organizations available on the USCCB website: *www.usccb.org*

Another program by the USCCB is called Amistad. The Amistad Movement seeks to empower immigrants, in at- risk communities, with the educational tools to protect their own community members from falling victim to human trafficking. The Amistad Movement has four goals:

1. The empowerment of the immigrants themselves through comprehensive workshops and trainings
2. The formation of peer educators from within the community to continue raising awareness with our support
3. The building of trust between the immigrant community and law enforcement
4. Coalition-building that enables the immigrant community to reach across and benefit from local NGO agencies and government programs working against human trafficking *(usccb.org)*

Available Amistad educational resources include:

- Trafficking 101
- How to Conduct Outreach
- Domestic Servitude (Labor)
- Trafficking in the Agriculture Industry (Labor)
- Trafficking in the Service Industry (Labor)
- Sex Trafficking
- Being an Ethical Employer and Consumer
- Coalition Building/Community Support/Integration of Survivors
- Human Trafficking Prevention Training for Parents
- Human Trafficking Prevention Training for Youth
  (usccb.org)

End Human Trafficking- Catholic Resources

THE CATHOLIC CHURCH AND HUMAN TRAFFICKING POLICY
HTTP://WWW.USCCB.ORG/ABOUT/MIGRATION-POLICY/THE-
CATHOLIC-CHURCH-AND-HUMAN-TRAFFICKING-
POLICY.CFM

U.S. Catholic Sisters Against Human Trafficking: Ending
Slavery is Everyone's Work
http://www.sistersagainsttrafficking.org/

COALITIONS OF CATHOLIC ORGANIZATIONS AGAINST HUMAN
TRAFFICKING http://www.usccb.org/about/anti-trafficking-
program/coalition-of-catholic-organizations-against-
human-trafficking.cfm

USCCB Anti-Trafficking Program
http://www.usccb.org/about/anti-trafficking-program/
USCCB Trafficking Victim Assistance Program
http://www.usccb.org/about/anti-trafficking-
program/mrstvap.cfm

USCCB Become a Shephard Toolkit
http://www.usccb.org/about/anti-trafficking-
program/become-a-shepherd-tool-kit.cfm

INTERNATIONAL DAY OF PRAYER AND AWARENESS AGAINST
HUMAN TRAFFICKING
HTTP://WWW.USCCB.ORG/ABOUT/ANTI-TRAFFICKING-
PROGRAM/DAY-OF-PRAYER.CFM

Stop Slavery and Human Trafficking | CRS - Catholic Relief
Services http://www.crs.org/get-involved/learn/slavery-
and-human-trafficking

Human Trafficking- Catholic Health Association
https://www.chausa.org/human-trafficking/how-to-help-
identify-victims

Statement on Human Trafficking-Vatican
http://www.vatican.va/roman_curia/pontifical_councils/ migrants/pom2007- 105/rc_pc_migrants_pom105_statementhuman- barnes.html

Franciscan Action Network
https://franciscanaction.org/issues/human-trafficking

Catholic Social Teaching on Human Trafficking – Intercommunity Peace & Justice Center
http://www.ipjc.org/links/HumanTraffickingAndCST.pdf

Caring for the Vulnerable - Catholic Charities USA
https://catholiccharitiesusa.org/efforts/caring-for-the- vulnerable

Catholics for Peace and Justice
http://www.catholicsforpeaceandjustice.org/advocacy- 2/human-trafficking

Christian Books on Human Trafficking

Finding Our Way Through the Traffick: Navigating the Complexities of a Christian Response to Sexual Exploitation and Trafficking, by Crawford & Miles

In Our Backyard, by Nita Belles

Setting the Captives Free, by Marion L.S. Carson

Stop the Traffick: Sister Mary Bad Habit Adventures, by George Patterson, OCP

Our Cross to Bear, by Charles Linhart

Numerous other good books, autobiographies, and films on Human Trafficking are available through book retailers and online.

Search "Human Trafficking"

Afterword

As I learn more about this horrific crime against humanity, I often wonder, where is the outcry? How can these crimes persist?

Despite the growing awareness and education available, human trafficking continues to be the fastest growing criminal industry in the world. It is long past time for us as Christians to see the reality of human trafficking and to respond to the cries of its victims. Whether it be a child working in the sugar cane fields, or a teen forced into prostitution, boys and men enslaved on fishing boats, or a servant in a neighbor's home, we must open our eyes to those most vulnerable and at-risk, and join forces with all faiths to combat this horrific crime. Human trafficking continues to be prevalent for the simple fact that the demand exists and the profits are high. Unlike a drug, which is used only once, a human can be used over and over again. And traffickers grow richer and richer from the *horror* of their victim's abuses.

What can we, as every day citizens, do to help end human trafficking?

It's a question that weighs on many of us as we hear more about the prevalence of slavery in our world, and even in our home towns. One way is through education. Many victims of trafficking come through our travel routes, health care systems, foster care homes, schools and even prisons, and yet are rarely properly identified as a victim of human trafficking. Learn the signs that identify victims and report observations to the local police or the national human trafficking hotline number:
(1-888-373-7888).

Educate children about the risks of internet grooming by predators and traffickers, provide training for all outreach and service providers, health workers, school and transportation staff that may be the eyes for catching this criminal activity. We also need to change our attitude and mindset about moral activity so that we teach, reflect and act with goodness in all places and times. Support families who are struggling, and ensure that they are unified and at peace so that no one is outcast to danger.

Secondly, we can either promote or combat human trafficking through our choices of the products and services we consume. Human trafficking is a supply chain issue and consumers are seeking more information and transparency on labeling regarding the treatment of workers in the processing and manufacturing of products. Think about the hands that created your clothes or harvested your food. Were they children? Did they suffer? Were they paid fairly? In order to produce goods cheaply and quickly, companies may be sacrificing ethical measures to meet the demand for cheaper products. When consumers choose 'fair trade' or 'ethically sourced products', they create demand for products that do not involve child labor, harsh conditions and unfair wages that contribute to poverty, and instead support the workers while providing safe, humane and sustainable employment. Fair and ethical trade policies as well as social responsibility reports are becoming more available to the consumer seeking transparency from the manufacturers regarding the details of origin and processing of products, and the impact on the workers, the community and the environment. As conscientious consumers, we have the power to support greater causes through our purchases. Read labels, ask questions, find out how your favorite goods and products are made and demand justice and dignity for all those that work behind the label.

Finally, in order to keep the most vulnerable people safe from the reaches of human traffickers, we must promote effective criminal justice systems by calling for tougher laws, stricter penalties, and creation of more effective strategies against those who exploit others and work to conceal their crimes. Become informed on politicians' platforms, and petition for support of ending trafficking. Only if the community is active and resolved to advocate for this voiceless population, will effective change begin to occur to end these horrific acts of violence and abuse against our children, brothers and sisters.

Kara Griffin
Associate lay member, US Catholic Sisters Against Human Trafficking
Member, Charlotte Diocese Anti-Human Trafficking Task Force
St. Matthew Green Team/Peace & Social Justice Ministry

REFERENCES

Alpert, E.J., Ahn R., Albright E., Purcell G., Burke T.F., Marcias-Konstantopoulos, W.L. Human Trafficking: Guidebook on Identification, Assessment, and Response in the Health Care Setting. MGH Human Trafficking Initiative, Division of Global Health and Human Rights, Department of Emergency Medicine, Massachusetts General Hospital, Boston, and Committee on Violence Prevention, Massachusetts Medical Society. Waltham, MA. September 2014.

Blai, Adam. Religious Demonology from a Catholic Perspective. http://www.religiousdemonology.com/books.html (Also see Cynthia B. Hunt MD and Monsignor John Esseff demonology)

Blue Heart Campaign. United Nations. https://www.unodc.org/blueheart/

Catholic Health Initiatives. (2016) Addressing Human Trafficking in the Health Care System. Retrieved from www.catholihealthinitiatives.org

Child Sex-Trafficking Recognition, Intervention, and Referral: An Educational Framework for the Development of Healthcare Provider Education Programs. Cathy L. Miller, Gloria Duke, and Sally Northam. Journal of Human Trafficking, Vol. 2, Issue 3, 2016.

Chisolm-Straker, M., Cossio, T. & Richardsom, L. (2013). Combating Slavery in the 21st Century: The Role of Emergency Medicine, Journal of Healthcare for Poor and Underserved, Volume 23, Number 3, 980-987. Johns Hopkins University Press, Retrieved from ProjectMuse database.

Council on Social Work Education, Center for Diversity and Economic and Social Justice. Refugees, Immigrants and Migrants, Education Resources. www.cswe.org

Coverdale, Gordon, Nguyen, Salami (eds). (2018). Psychiatry's Role in the Management of Human Trafficking Victims. Retrieved from ResearchGate April 3, 2019.

Dalsfoist, K. (2016, September 5). Tagging device found in human trafficking victim. San Francisco Globe. www.sfglobe.com

Diagnostic and Statistical Manual of Mental Disorders, Fifth Edition, (Copyright 2013). American Psychiatric Association.

ecpat International. www.ecpat.org

HEAL Trafficking. Health, Advocacy, Education, Linkage: Because Human Trafficking is a Public Health Issues. (2016, July/August) Retrieved https://healtrafficking.org

Kerr, P.L. (2016) Treating Trauma in the Context of Human Trafficking: Intersections of Psychological, Social, and Cultural Factors. (pp. 199–221).

Laboratory to Comabt Human Trafficking (LCHT) combathumantrafficking.org [Core Value: Interdisciplinary Response]

Le, P. D. (2017). "Reconstructing a Sense of Self" Trauma and Coping Among Returned Women Survivors of Human Trafficking in Vietnam. *Qualitative health research*, *27*(4), 509–519.

National Association of Social Workers. (2015). National Association of Social Workers Code of Ethics. Washington, DC: NASW Press.

Navajits, L. Seeking Sagety: A Treatment Manual for PTSD. (2012). New York, NY: Guilford Press.

Not My Life. (2011). Film. Directed by Robert Bilheimer.

Riebel, T. (2016). Healthcare Screening of Human Trafficking: Qualitative Social Assessment. DOI 10.13140

Shafoori, Caspi & Smith (eds.). International Perspectives on Traumatic Stress: Theory, Access, and Mental Health Services. Hauppauge, NY: Nova Science Publishers.

TESI-C Child Trauma Screening https://www.ptsd.va.gov/professional/assessment/child/tesi.asp

There is H.O.P.E. for Me, Inc. wwwthereishopeforme.org to donate and contact Dr. Kat for Speaking: www.DrKat.net

United States Conference of Catholic Bishops (USCCB). (2019). Retrieved from http://www.usccb.org/about/anti-trafficking-program/upload/SHEPHERD-Leader-Guide.pdf

United States Department of Health and Human Services, Office of Administration of Children & Families. SOAR to Health and Wellness Training. Retrieved April 3, 2019. www.acf.hhs.gov/endtrafficking/initiatives/soar

Williamson, Dutch & Clawson. Evidence-Based Mental Health Treatment for Victims of Human Trafficking. Published in Treating the Hidden Wounds: Trauma Treatment and Mental Health Recovery for Victims of Human Trafficking, by U.S. Department of Health and Human Services, Office of the Assistant Secretary for Planning and Evaluation. Retrieved August 23, 2016 from https://aspe.hhs.gov

Williamson, Erin, Nicole M. Dutch, and Heather Clawson. Medical Treatment of Victims of Sexual Assault and Domestic Violence and Its Applicability to Victims of Human Trafficking. U.S. Dept. of Health and Human Services (2010), available at http://aspe.hhs.gov/hsp/07/HumanTrafficking/SADV/index.pdf.

St. Josephine M. Bakhita

St. Josephine Bakhita, you were sold into slavery as a child and endured untold hardship and suffering. Once liberated from your physical enslavement, you found true redemption in your encounter with Christ and his Church.

O St. Bakhita, assist all those who are trapped in a state of slavery; Intercede with God on their behalf so that they will be released from their chains of captivity.
Those whom man enslaves, let God set free.

Provide comfort to survivors of slavery and let them look to you as an example of hope and faith. Help all survivors find healing from their wounds.

We ask for your prayers and intercessions for those enslaved among us. Amen.

United States Conference of Catholic Bishops

St. Michael the Archangel,
defend them in battle.
Be their protection against the
wickedness and snares of the devil.
May God rebuke them, we humbly
pray; and do thou, O Prince of the
Heavenly host, by the power of God,
cast into hell, Satan and all the evil
spirits, who prowl about the world
seeking the ruin of souls. Amen

Jesus Christ, consubstantial with the Father
and the Holy Spirit;
through Him all things were made.

Holy Spirit, terror of demons, save these victims.

The resource information presented in this book is of the public domain
and written according to current data at the time of compilation.
Information may change, be removed, or evolve over time.

www.ingramcontent.com/pod-product-compliance
Lightning Source LLC
Chambersburg PA
CBHW030911180526
45163CB00004B/1787